searching for Science

Elihu Blotnick

*fire*fall™

San Francisco
Southerness

first edition: April 2019

print: 9781939434609
audio: 9781939434616
© 2018-2019 Elihu Blotnick
all rights reserved
firefallmedia.com
prize@att.net

LIBRARY OF CONGRESS CATALOGING-IN-PUBLICATION DATA

Names: Blotnick, Elihu, author.
Title: Searching for science / Elihu Blotnick.
Description: First edition. | San Francisco : Firefall, 2019. | "Firefall
  Originals."
Identifiers: LCCN 2018042550 | ISBN 9781939434609 (pbk.)
Classification: LCC PS3552.L683 A6 2019 | DDC 811/.6--dc23
LC record available at https://lccn.loc.gov/2018042550

Stamp of Approval
single digit celebrations
(1st findings)

*T*ime protests. Tethered to truth, science
is as safe as an Uzbek trenchcoat with moth-
wing sleeves. It's not waiting for evolution
to cocoon it again. Temporarily it will
equilibrate, but change is a flutter away,
and end-death, if no buddha is actively
teaching in the world. A gold bow with
one silver arrow, so powerful it can lead
to enlightenment in a single lifetime;
a rain-shine green that glows in its own
shade; moist oxygen-rich air to breathe
life back to its roots — our ark needs
no wheel house, unless a pair denies
its partners. Each species depends
on devious defenses. Weapon eyes
blind ego realities, and vice versa.
Collider-bought light survives only
a pico-second to flash us its state.
Earth predictably has nowhere to go,
but we do, with our pluripotent stem
cells. From the water's point of view,
anomaly unicycles into geo-salvation,

by coupling to itself; flushing oceans
don't guard their polluted identity.
Up-lifting souls on the sweep hand
of science isn't as easy as the pulpit
preaches. Rose-touched windows give
face to undiluted contaminant wrath.

———————————————➤

(Common sense has changed
again. Coding backward or up-
side down are in. Laws against
free refills are where the action
is. Disruption seduces with its
simplicity. Chaos is a green
light turning. Bit of a muddle,
isn't it? Einstein's analogies
future into the past: endless
elevators, super-symmetric,
triple our visual brains in
semantic space. New names
are everywhere. Limitation
is self-contradictory. Reverse
engineering Earth is the first
test of our dreams, but no
more than a mathematical
notion. Our ban on cotton
swabs has no equal sign. All
theorems need a proof, yes,
but boron blue diamonds dive

too deeply for quick reach, yet six
guardians of our genome claim
the media rights of meteorites.
All world class wand wielders
day to day, one keeps her last
husband's ashes in her glove
box, waiting two years now for
the chance to dump him into
the bay. Style for her distinct
vintage is a layered mis-match.
The real deal is always a classic.

——————————————➤

*C*hemistry committees need bio-prizes
or a bill of rights, loud and proud as a baby,
to force feed our being into white fire,
that grows without ash or extinction.
Calling all Earth to vote for a future you
can glow by. The League of Less invites
you to discover more than life: chemistry,
naturally uncaged: mindful mending,
the embrace of sympathetic science:
self-evident choice without chance,
beehive brave: behaved scintillations
creating new waves from dim light,
air-years from now, near as relativity.
Science sings true, when consenting
and focused, intent on survival against
blustery, lurid denial: pop lore, snotty,
knotted clusters of vibrant discomfort.
Oh, yes, I saw an eagle today weathering
in time on the fission tree, a plasma pump
in its claws, in a simple, sentient symphony
scientized, oxygenated, and newly nesting,

before fiercely rising, with the speed
of celebrity, in clapping winds to a future,
symbolic of seismic wonder. Do friend
Earth, to save naturally; horizons are end-
less, as urgent as gravity. Answer the dark
by living light; biology is chemistry alive,
with humans inside, in suspense. Science
saves us from ourselves, if we trust it twice.

———————————————➤

We can blind to the sky and long
for dark riches on Earth but even
eclipsed the sun still surrounds
the moon. Or spend out on travel
to the space port. Day trips to Mars
are still far off. Paws of unkind laws
imprint us. Science survives itself;
Dracula orchids and purple anemone
don't, being as unrare as we are.
You, I, and the universe insist
our grass is leafier, waves taller,
nights whiter, ice bluer than thou.
Surprise is everywhere. Two-headed
constants divide us. Data matters
more, dark matter most. What we
don't know is two stones with one
bird split between them. The ark
of life floats free on fragile slurry.
The existential lace of reality
under an atomic force microscope,
no knitting necessary, is finely
graced. Look first, then go where
the science is. Decide Sapphire

or Cyanide. Apricot or Arsenic?
Emerald or Argon? Ski the Rain-
bow. Nuance nature, to graze
on dry gold, eroded. Valleys rise,
fall. Wash never with mercury.
Only safety valves survive freezing
steam, boiling ice, acid fermentings.
Even the door knob at the day
care center isn't germ free.

---------------------------------➤

W*ith* three layer brains, shaped by four alphabet-
block genes, and chromosomes enough to choke
a gopher, a magnitude or six more and we might
reset as aliens ourselves, crippled and deformed,
or alive as never before, newly born into our next
gen being, fully awed by our acappella excellence,
independent of Earth, so it seems. Life serialized,
science realized, eco-idealized, freshly entangled
all in the other. Geo-sci echo-ups. That's theory
to a reality. If energy is proportional to momentum
squared, chance is choice by itself now. Let helio-
physicists delete their cross-eyed morality plays,
rid their orbits of sophomoric static, and cheer
us in arcs of enthusiasm, of the unblemished reach
they will own, by imaging a seventh born star in 3-D.
Abruptly then, three leads to four brains, and four
letter genes leap past six levels of evolution. Suddenly
we are fireflies brighter than Venus, and life is lined
with penguin down. We give ourselves permission
to launch, for space walks. Our sky is ground. Time
retreats. Swans drown in deep night, in the patience
of space. Not all particles are created equal. As we
know, dark matter can not jump rope, yet when

photon lands on an atom it gains mass. Absolute zero, our hero, subsumes all. Beware — science is here, the sum of all outcomes; our plan for our universe is a balance of beliefs, un-triple-jointed. Lift off is T minus ten. You have one week, forever, to free-associate with our geo-sphere.

———————————————➤

I couldn't resist the delicious
temptation of her stinging
stare, the chance to thrill,
taste, glory in the fecund
wonder of her, to see her
tensing as the beginning
of time, a bee on the moon
in an infra-red beam, clearly
signalling heat. I'm the cold
master of attitude and orbit
control with 1000 theories
a year from one telescope,
a doctor ever bidding on
patients closer to my
modality, a transient with
4x4x4 tables to map prime
sequences onto amino acids.
Kansas repealed the laws
of science now a bible ago.
Global warming channels
are unpleased. The bullet-
proof uterus may soon be
the fashion; we'll still spell-

check a genius visa, video
sheep on a trampoline in
deep curiosity, and imprint
ourselves daily to flash-
heal the real, repeatedly.
Half-deployed sun-shields
are like spawned-out squids
dying on arrival. Addiction
feeds itself, stuns our sun red.
Only science hears a buzz-win.

—————————————→

*S*cience wins; science survives wild-
fires, hurricanes, its own laws, always.
It's an outcome of focused accident.
an energizing brain, ranging from 7°
kelvin up to fusion, a set volatility
never satisfied, a fractured search
for equal signs, happy to surprise it-
self playing dice on splintered wood.
Simple really, science prevails first,
last, and always. Science bets on snake
eyes in a poker face. No one need slink
away from the tables. Truth is self-
healing. Science is right till found
wrong and wrong till proven right.
Science wins without selling itself.
Science soaks up the sun, absorbs
the night ice. Only a greedy sales-
force gushes forth about the science
of awesomeness, silly putty, and, yes,
 two-headed mice. Science, abused
by chain store freak shows, creates tiny
spies, unspellable medicines, ingredients
I can't pronounce. Casual testimonials

never tempt us to believe: Didn't the US
end today? Nostradamus came to New
York, defeated, predictions off. Sparks,
from burning astronaut pants ejected
from the shuttle, ignite a NASA-speak
release. We're a space-faring species;
another galaxy is where we'll find the
truth about Earth we missed here.
Science is our universe; even if rogue
planets in damage mode drift our way.

—————————⟶

Which side of the seesaw do you want?
Hefting colorful feathers or iron cubes;
with the ungoverned or a governess?
Up, down, faster and faster, forecasts
race under wedding rice. Uber versus
alles, mucus like rubber cement,
dries before we stick. The redundant
hunt, rabid enthusiasm, temporary
permanence, let the superstitions
of science, like Iranian death chants,
'Death to death!' curse scrolls, rail
out, refusing to believe in light speed
limits. Those scrawlings are extra panels
in an odd fan not stirring our cultural
air. In a fractured universe, enemies
altar up, yet our newly-bred frowners
still catch a breath of laughter: with yet
another "spy ring of squirrels" busted.
Ship the lost to Mars; tolerated habitués,
they've a chance, their survival skills
unstunted. Which future, whose past,
what present fits is too much to ask.
A seesaw of left and right-handed

molecules inclines to the poison side,
not following its link to self-regulating
laws. Pine nuts and nuggets don't level
the scale well. Click streams drown us
down-river. Upsets are inevitable, when
impulse prevails. We need magnetic cages -
holding in the solar-haired. Yet, death
is not intimidated, nor is the CO_2 flared.
We are born binaries, but unbalanced,
ever in need of full natal cloud worlds.

The experiment in human projection —
orbiting the Milky Way, piercing the wall
of hydrogen at the heliosphere, to sort
out ultra-faint dwarf galaxies, translates
back into star apps for googly commuters
and hexacopter goggles for tattooed eyes:
monocults framed to exclude the barking
nagging darking for more — even ants
intimidate the tinier. The NSF compass
in the genetic pilothouse encircles
kilo-nova events, as earthworm minds
drone on about unpuritan future ex-
wives. Mutations make genes useless
most times. Clowns and comediennes
working the parking lots alone think
they show better, but science is on its
own side; ours too if we chase challenges,
not leaders; meet comets head on;
spin with spiral galaxies; dawn with
the day; pick new horizons over banana
wheels of foam-loathing. Sure, science
is the old maid in the deck of longing,
denied spin-coupling, yet embracing

us all. Our renaissance in space turns
our bones to jelly. Molding, Odessa
seeded rye chokes the brains of the
left behind. With and without fail,
science is the future of everything.
Herd immunity is a major myth.
Science is the only survival strategy
to drift toward, dance with, cling to,
awe through without fright. A mind
declined has an instant decay rate.
Resonate up with an earthly meta-
physical dashboard, to light your fate.

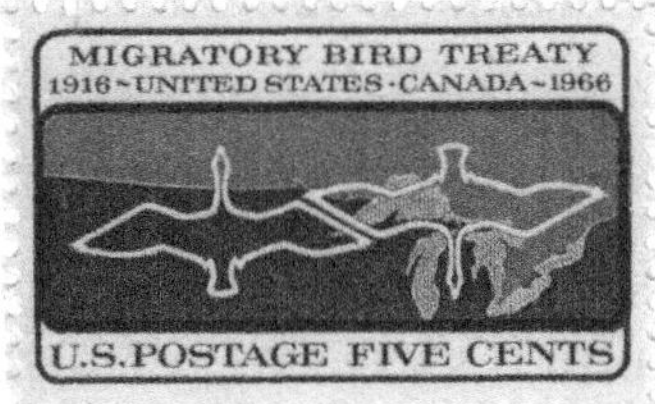

—————————————→

Sanctuary science exists in Cities of Truth,
labs in the thin dry air of desert mountains,
at campuses where the social doesn't hold
sway; a subsidized supermarket is only
a bike ride away, here no one overdoses
but on essential equations, conceptual
models, or big theory. Who plays, who
plows, whose sheep block the road just
two miles west are questions unanswered,
or moot curls of no point. Only knowing
engraves the currency of conversation.
Celebration is success, discovery, the next
test. The atom sings solo, the molecules
in chorus, but up the corridor is a collider
that creates its own light, by igniting tiny
suns in near-zero space. Security is so
tight, the crown jewels are invisible.
Even the predictable is retested, re-
fined and enshrined to subliminal. No
silhouettes walk the halls, yet engineers
rush. Science is self-healing, yet rupture is
the goal, not rapture. The playlist is nature
only. Freely pin this story to its conclusion.

No pregnancy payments are needed,
so fly blueprints on lunch-napkins
to the northern lights, and ride uni-
cycles into the future to taste a new
spice. Let honeymoon babies with
their fingertips grip the green life,
and re-recycle. No systems are here
to fight, own, or repair; and no first
drafts, only unedited truths; no fuse
but a switch to flip skyward at last.

*W*ashington is being blind-sided
again. Astronaut senators, that's
one sight. Ordinaire scientists
running for high office another,
but it's in the mirror now, back to
pragmatic, owning power to trillion
dollar problems to death: Mars first,
then misguided-martyrs with flaming
shoes. But show a candidate hacking
a voting machine in Times Square
with no help and enable your winner.
Burning fingers off on hot buttons hurts
less than losing. Passport protected,
we have on the easel a famined Turkey
and crippled Iran. Endangered species,
the padded homeless, and refugee
camps downtown are solveable,
maybe. Yet a politician is an ostrich
egg boiled in a plastic bag, a problem
in topology, semi-cylindrical, overly
stubby, but ever-visible. Worse on
the nerves are the slogans repeated
but needed that deter the gifted

enough to pen a haiku in view of
battle sets. Science solves, saves,
satisfies, and suggests perfection,
but it's incremental. Political pride
competes with itself in public. Still
a new wing to the White House will
want its flag green and translucent
to set a standard far from the dark
webs. Who's saluting? DC dropped
to its knees is a many-limbed ritual.
We need a national quantum strategy
really to revolutionize the world,
and leapers to lift the workforce
future, to shape the next meta-
physical reveal. The drifting kelp
of culture on the swift currents
of science will vote in the right.

————————————➤

*G*ravity without matter is a mis-
calculation. Formulae are self-
correcting after all. Run life back-
wards and it's still in a forward
framework. Science is a need
to solve for all unknowns, but
distracting herself with a bubble
wand from a thinly sliced lotus
root leaves a soapy smirk on
the diet doctor's lips. The tidal
flow of blood reveals a leaking
valve and a split signal missing
a beat too. Yes, dueling uteri
twitch in their trap. Ballistic sub-
atomic particles and frustrated
molecules join to alternate ions.
Species self-interest isn't simple.
Hairpin DNA is only one i-motif.
The dynamic still has checkpoints.
Petrochemical hair coloring may
wash out, but the original can't
be natural anew. If we play with
the pH of the dye, we get blue,

gold, or chrome. Yeast is a good
engineer too. Little is reversible.
Trying to steal from time is too
like swiping wet underwear from
a laundromat and finding it over-
bleached, shredding, ill-fitting,
an absurdly futile endeavor, but
we accept some risks, if the sense
of wrong feels right. Abandoned,
setting out with bad maps, we will
still potentiate even on all-world
menopause day, now celebrated.

$\longrightarrow$

S*cience* doesn't play it safe or need to.
We bow to ruling codes, with fissionable
plutonium, or over-heat the ionosphere
competing with the Russians in fucking
with the weather. Failing at that, we fall
back on neutron alternatives. If we don't
experiment, hell, others will and not give
us the antidote. The unthinkable is here
now. It's self-funding. With or without
us, the future is a flip of a coin, whether
we call it tails or heads or edge up. Grafts
and hybrids are valid gifts of being, if they
survive new settings. Native plants do not
do well in rifling weather. They seed north.
Fish chase the gulf stream, and bees die
off. Birds won't migrate. We still target each
other on the wing. We don't need new senses
to know ourselves. Sharp science transforms
into chemistry college spinoffs. The cost
isn't high but narrowing ladder steps trip
us into falling. Yes, we start smart, become
half-smart, deluded, and delusional lastly
down in the dust, ringed by knockoffs.

Newly original, we don't ache to save face
as in China, or wear smog masks, or re-copy
them copying us, but even if we out-wit
the weather with orbital leaps, it's baby-
stepping. We will free the China Sea,
if need be. All is possible by accepting
the unintended. Morality is a liquid
trapped in a liquid, a gummi worm
meowing, mauing for more in swarms
of obedience. Super science is best alone,
but single-use needs only one twist of
the last setting. Press any button now.
Micro-climates are global triggers.

————————————➔

The science of selling science is
not science at all. It's a closed loop,
feedback-driven, self-defined;
the uniform is pressed too neatly
into a confining solid wall. Unified
dress-codes restrict priorities better
improvised. Science is an onion
of a word, after all, layered, seeping
into the air till we see by weeping.
Cooking, it slowly caramelizes,
tantalizes, and releases its syrup
of truth and taste for conviction's
sting, quantifiably, in a test of full
control. Simple, shareable bee-
feeding flowers lovingly tended
are our color guide, the clue to
lemon, betel, and curry leaves,
red beans, green garbanzos, opo
squash, ramps, matcha, hibiscus,
and poisonous milk. Mass spec
reveals the subtleties identifying
Indian yellow, mummy brown,

and nanotube black, as well
as clear uranium green glass.
A pigment is a fingerprint,
a hold on life, rigorous, real,
and repeatable: an exact science.
We then can discover vallinoid
skin receptors that detect x-rays
and scientists as oh-so scrupulous
connoisseurs. Authority is finely
textured, not a focus group, whose
seeds will burst and eggs turn to
ash un-surviving a sale-hot Earth.

⟶

Science for science's sake is a jungle
gym of possibilities; basic research
is a tree of rare chance to climb.
Saving science from the masses
helps; not needing to nets us more.
The public only want the smooth
fruit anyway, not a wrinkled mind-
set; their disdain is a tossed-off
sneer. Let them be denied. Give
us back our free pass, take away
theirs, even if synapses missing
show up in ink as bellowing
sadistic outrage and shaming
growls. Yes, news pangs with
hunger without a no-cost
source. This is called politics
by a party willing to throw up
its future. Deserts are always
deserted to them. Still, suited
in style or walled off, branching
into shadows across a clean slate,
we leap, fly down, and swim deep.
Post-docking or tenure tracking,

visitor or rotator, researching or
applying, the material science is
much the same, till a sky-diving
dentist joins us from trial sleep,
and the curse of the black granite
sarcophagus lifts to her rhythmed
sighs. With chalk bones and blood-
less marrow, flanked by glow strips
and rolling white-outs, she asks,
with a smile through charcoal teeth,
"Rist? Rism? Pissed or prismed?"
Orthogonal matrices in the making
surprise our solid stable reach.
A black hole still sits dead center.

———————————————➤

S*trict* science isn't tolerant;
an RV to Mars and a coal tar train
are one and the same in North
Korea. We have antidotes to
neutralize the minimum lethal
dose of back-sliding. The single
frequency mind lacks capacity
to endanger itself. Gumbo with-
out okra isn't as thick as the fat
man is thin. Putt butter and ball
hero to himself, he's a good send
to renegotiate our place in his
universe, with aliens as aggrieved
and aggravating. What a table
breaking meeting that might be.
Doubling back isn't easy when
you've gone too far. Our invisible
weapons are safely unseen. If
we hear bombs dropping, then
even our shadows can't take our
place, no, but we'll start all over
hopefully, or someone else
will, if it's not too late. Green

riots above dry aquifers on
farms in new deserts. Eco-
chemistry owes thanks to itself.
He who forgot to engage a stop
signal, who rubbed the ruby
of reality the wrong way in spite,
brought the book of the dead
back to life, and the sun down
to Earth, deserves some credit.
He, like strict science, has few
limits on his own hot fears
of burning up the atmosphere.
Fueling the thrust of a 10-g
ego-ecologist eco-grumbling
is pure play. Our rule will wait.

—————————————————➤

Speak up, science; act out
as never before, or turbulent
simulations taking place in
the brainport will go black
before the flash-healing. Neo,
meta, and super still predict
science is safe except when
it isn't. Science sacrifices not
one thing in its way. Nor must
we, as early adapters, in every
language, with sixty consonants
and two vowels, if need be now.
Electing laws worldwide simply
won't work. They must first mold
in the soil, rise in the air, replace
routine with rose water and free
music. Cultural magic, the tricks
of pseudo-science also, whatever
serves survival is welcome too.
All is allowed, even 3D printed
food and a ban imposed with-
out debate on HFCs in frozen-
yogurt dispensers. If this grates,

like inhibitors, to create slow-motion,
or NGOs with claws threatening
discomfort at a late night resort,
we must persist to infinity.
Yes, if accidents surprise, then
science too, through us, immune
in its equilibrium, must scream
out in prophetic squeals and harsh
thunder. Always it's up to us alone
to be ambassadors of the future
state of grace. The architecture
is incidental. Policy and planning
borders aside, we speak Earth!

*F*lavors and fragrances, subtle
and strong, real or artificial
do little wrong to our taste
for life. Color-coded levels
of trust are counter-intuitive
though. Less is set in human
cycles than we like to believe.
Resolving a racemic mixture
needs only polarizing magnets
to separate the ups and downs,
the effective from a side effect.
Yet, our senses wake scented
memories from sleep to meet
us in repeat. As journalists will
say, "The news is ever the same,
we just change the words," but
science reveals more than four
story lines and less than nine
laws to the schema of things.
All claims of a pin with no
head, a needle with no eye
are ill-defined. Then too,
recalculating one watt is not

a waste of time. Like π we
never reach the end limit
of precision. Constants lay
entwined in wedding beds
waiting for a timeless baby.
Wonder is a life's work. Scale
up a custom reactor and find
human truth is evanescent,
yet suffocating collectively;
universal reality is ineffable,
odoriferous, tangy, ballistic.

C*an* fractions exist without whole numbers?
Do family dinners explain our brain size?
Is a particle's pico-life proof of its existence?
Are femto-misfolds the reason we're bipeds?
Must we start with the hypothetical always?
If we've an answer, why do they need to know?
How many suns must explode to make gold?
Does it matter which nanos create the nugget?
Why is the unanswerable everywhere now?
Is the number of cell deaths fixed?
How many notches make a niche?
Do states of being divide all life forms evenly?
Why really did worms learn to walk?
Bread-loaves — cube-sats — doves —
What's in a name when they're all the same?
Where do we go from here? Why is that?
What if all formulae are wrong or don't fully fit?
Is mathematics the only secret of the universe?
How many comet hits can a planet take?
Can our life exist without H, O, and C combined?
Will a silicon-sulfur based humanoid see us too?
Must we always challenge the obvious?
Is nothing a foregone conclusion, even in part?

Are the rules of science too leveling and limiting?
Is climbing a learning curve quicker in a vacuum?
Are bacteria acting in concert an ajacency or an all?
Is a virus a vector, an opportunist, or our next life?
When bacteria go bad, is the victim the trigger?
Should we outlaw the extra-dimensional?
If three right turns don't turn us left, what's wrong?
As gravity is so weak, why aren't we stronger?
What's our next bloody broken-bat challenge?
Is dark matter a dream of disproportion?
Step up with your science, take the next
swing, fill the six gloves in the far seats!

*S*cience CEOs are rare, yet full of fresh
blood and crossing barriers, their success
is served. Being a pressure-cook chief
means reducing 39-step processes to one
team effort, recruited up from the ranks
of eco-in-laws, and green-washing
the brain gain openly, to create new
viral vectors and safe fungi to sell, true.
Still, should share prices fall as trials
fail, the public will see villainy in each
shadow and death heads on all front
doors. Story trumps the quadratic
though. So, lipstick line the universe;
then mine bacterial genomes. Let
wasps eat the stars, then say the moon
reversed direction last night; reveal
the butterfly in the bark, to watch
the pendulum swing in your favor.
Primitive science is magic, cave
paintings for all, but alter a micro-shift
from cotton to carbon to composites
reinforced with glass thread; uncurl
up from wool to nylon-6 spinning

wheels and back to bio-based indigo,
notwithstanding tariff extracts
and royal die-offs; and seduce
other CEO pools to sight new
tasks in your quantum cross-
hairs: the job has just begun.
Between barium and hafnium
are plenty of half-wits and morons
in reactive ion states, out-averaged.
An all-in-one bomb, from a stealth
fighter, still pilots brightness best.

———————————➤

Breaking the big
claims down
to small clumps
works once
in a while.
Orbits are lost
in tradeoffs, if
the fit is foam,
or newly toxic.
Five eyes watching
turn unfamiliar.
Sectors are shards
only, unlinked, unable
to re-attach unless
new proteins offer
true futures,
or re-grant old powers
non-existent in nature
till now. So, sign up
for a chromosome
or a sector of one,
and re-write the billions
of base-pairs to recode

eternity's best questions.
Stitch string theory
through the needle
of time and reward
yourself with knowing
how much more you
don't know and never
will, but life may give
you another chance.
Take it in the fabric.

———————————————➤

"Is the sun coming out today?"
asked the elder on the bus,
before the dawn in question
on her way to the art
and science center where
nothing is taught
but nature most mornings.
That day though, the ring
of fire erupted north
and south of here
and cracked the street
so deeply she never
crossed it again.
The Neptune Society
followed up, while across
the bay eleven barges
of polyvert spilled
perversely in the tidal
back-wash. The particles
layered the fog, forbidding
it to lift again, but the sun
waited impatiently

everywhere else while
the Parker Solar Probe
came back as a line dance
to *California Dreaming*
and the new Art & Science
center closed, though
science is more seductive
by the day, not to mention
the night sky and the NASA
channel. Culture is shifting.
In its new mode, suns don't set
but Earth still spins, cringing.

$\longrightarrow$

A *so*-wrong knows-it-all,
braying, beats a slow seeker,
scripted humble, in cowardly
retreat, on his knees bowing
to the east, but any and all
excess must be checked or
it will outface itself as soon
as sunset. Glow-in-the-dark
water fights on TV elevate
the body count, but early-
stage immunity vs. graying
celebrity juries out in a court
fight. A Russian beard tax
won't weigh our scales again.
News will remain the Trumped-
up Hour, all segments leaking
bile, as we sip up to the next
election. Curiously, drunken
pilots and hospital nurses lead
the list of suicide prevention
calls, in San Francisco at least.
Are Russian hackers rehacks

from the NSA? Does it matter?
When firefalls light mountain
walls, safe shrines and soothing
sleep riddle the big cities. Trolls,
crushed under their own bridges,
sacrificed like lost mice in our
cyclotrons, cease to exist. Well-
cradled presidents, offering
uncertain substance and con-
fidence for all, try to enlist
us, but embalm their office.

Does an ancient universe
survive in ours, in part? Is it
detectable? If so, ours is not
the first of a kind. Hawking
points are simply random
noise. Really? Sure, the white
snow on our analog TVs
is from both this universe
and the last. Just before
our Big Bang, naught existed
but black holes gathering
to explode. Their massless
remains are the missing dark
matter of the universe: photon-
fragments. That's my theory.
If it's pure ghost to posit such
an earlier oneness, the guilty
pleasure still belongs to all
who claim it. Science is a part
of speech, a set of formulae
and facts, a modifier of method,
a sharp attitude. But loosely

defined, perverted in purpose,
misguided naively, malevolently,
or just out of wishful thinking in
a leap sideways, science may also
serve death. Language is precise
in its intention; its most fertile
soils grow a garden of still-lifes,
a hawk moth on a white orchid,
as well as a pit of green mambas.
Science fiction too tears up rules,
validates non-proofs of knowing.

———————————————➤

Science is branded
in to all ransoms handed
past. Dark web banking
in block-chains meters
up enough heat for two
oceans filled with plastic
fine-tuned to self-destruct
but failing as future fuel.
Inflows rise and over-blanket
the flux of burning breath.
The debt is inexplicable,
yet simmering, senseless,
blaming science for lost
fortune is mined market-
forced data from all two-
edged sales games played
for profit or comfort. Science
sells sin at the kid's counter:
magic science and the science
of magic raced up and down
the pillow face while the art
of science and the science

of art locked in embrace.
So in so said, this is selling
success twelve ways, yet
the counter isn't often licked
clean. Only Navy seals survive
their nursery rhymes. Science
needs similar weapons. A one-
planet species fails eventually
like a sales pitch that leaves
us street-wise and stone-deaf —
as science descends through all
brands in the gutters and run-offs.

————————————➤

*S*cience is a carrier wave;
the frequency is fixed
but can be redrawn to
pencil flat the artifacts, or
any arbitrary static inserts,
yet mouth noises appeal
to ears resonant at all
levels, who *will* life
into every wave length.
We are always born too
early to learn long term
outcomes of the next
iteration. A square mile
of silvered solar mirrors,
to a magnitude more,
is not unimaginable.
Science owns every
frequency but flavors
the most telling. Truth
is, if subtlety is lost to
flare, then melting down
silver trumpets frees

wide-open messages.
Any one-valve bugle
will do too, to wake
us, but our morning
march requires clear
urgency, the need to
hear and see simplicity
in the complex, or non-
trivial variables found
only in ears tone-deaf
to oscilloscope screens.

*S*cience is a law unto itself; its power
pervades us. Peer review staffs the feed.
We lack a common metabolism to agree
though. Each to his own mantra, as it
were. A neon carrot-blond with a long
drawn face thinks expectantly: 'If sun-
screen bleaches coral reefs, but blue
light penetrates my skin more deeply,
then intervening clinically is a crap shoot.'
Control is the healthy terror of doubt
not lost to error. There's no one to blame,
and no need for blame. Stone throwing
only skips the truth wildly. Every mind
floats in its own knowing. Do-it-yourself
politics create the enzymes for science
uprisings of unexpected constituencies,
of strange limbs batoning back to Earth.
The public, CO_2 brained, insists, "We're
ordinary," meaning 'Expect Nothing!'
Not slick, like inadvertent suicides
who park on the tracks and wave at
the trains to go around them. Or a flat-

bed driver who unloads by repeatedly
stopping short in reverse till the lumber
falls off at the job site. There's always
a way to use a law against itself. No
subtlety needed. Just split the legal
momentum, by alternate braking,
or contrary re-use. The primitive
has survived far longer than we
have; its power still renews us.

_______________________➤

*P*olarizing the paths of entanglement
limits the possibilities to two non-
identities, when, hell, a wheel of shades
across the compass rose is available,
in every color too. Not shadows, but
alternate substances are at play. Our
toys fix any and all positions in a sub-
atomic *Kama Sutra,* if we're inclined.
Naturally we are. An equivalence is
An Abattoir of Wonder, not yet penned.
Are we deluding ourselves, hoping
to leverage our manipulations into
partnership with the universe? Likely
so, but bleeding the leech of the blood
it stole is legal too. Precision-guidance
in a threat-environment requires long
range panning. Not all gold dust is equal.
Nor should we want it to be, but slapping
a sack down on the bar gets dynamite
attention still. Creating unbreakable
code is hardly worth the effort if just
one spy can attach a thumb stick

to the wrong hand. Life is quite bland
otherwise. We learn! "They're not
paying me enough," says the fighter
knowing he's about to lose, leaving
the ring two seconds after the first
bell. The winner can't gloat, with no
pain delivered. Yet, on any given day,
anyone can win. Wait! Do you have
to visualize an outcome to make it
happen? Yes, like cell-phone drama,
all is on the screen; the capacitance
of fingerprint ridges is still decisive.

$\longrightarrow$

*S*cience is a submarine
submerging all we do.
Science is a style, its own
point of view. Its search
for ghosts, hosted
by theory, nobly
predicting its autumn
harvest, a cornucopia
of constants, sprinkled
with spices, from every
field and port well met.
The saffron on our chicken
wings and greaseless links
comes in every red, like
the densest dying stars.
Even puny periscopes see
the universal barbecue,
while sea snakes
with their own sonar,
wait for us to pass.
We may drink oceans
of dissonance and still
know where we are,

or inhale plankton or
more elementary particles
taste-free in endless breath,
or stay under seemingly forever,
or surface to sand castles
or essential formulae at work.
The ocean is a lemonade,
five-stars, acidified, not
quite poison-free, yet.

———————————————————>

*M*indful mendings:

Trust truth, the sun,
and no one but —
The power is yours
to wind-surf with the whales;
to breach each storm wave;
to survive every aura of space.
If science is life, we can choose
to love it to death, or liberation.
Yet, a tweet in your face sours
the taste of whimsy. So, cut
restricting cables, blunt
the harpoon barbs; hammer
down walls; it's natural to grow
tall. The law of disorder will
not change the fate of a formula.
Send for Einstein: the science-
minded light up galaxies.
Keep models elegantly simple,
imperfect yet precise. The quirky
perversity of an energizing
frequency is the real song.
Every sensor is its own measure.

Liquid-nitrogen will freeze both
fears and nerves forever. Hard-
won, easily lost: high density ice,
a bouncing run, the clap-and-fling
of flapping wings. If the saturation
energy of an inverted pendulum
is not real to some, and awesome
to many, are the entangled aware
also they can be torqued differently?
Alpha equals have a reference signal:
red. white, and blue equal green.
The bigger wiggle is always so telling.
Rulers who rule all but themselves
will tune to neuro-bubble-tech
next; a night light still leads all
necessity. But how snakes sidewind
and sandfish swim arrow us into wide-
eyed surprise and brighter purpose.
Silently fate still wins, unless we step
in — with our flight of metaphysical
findings and a fistful of quantum suns.

Made in the USA
Monee, IL
07 July 2026